FUN WITH

Shadows and Reflections

BARBARA TAYLOR

KINGFISHER BOOKS

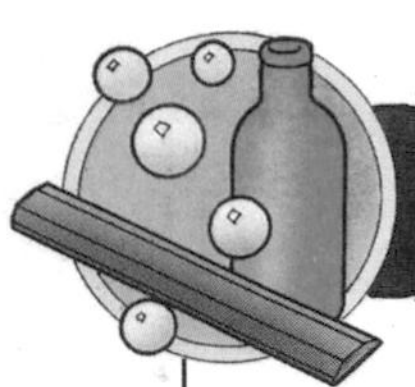

SHADOWS AND REFLECTIONS

In this book, you can discover how and why shadows form and take a closer look at the reflections in mirrors and other shiny materials.
The book is divided into five different topics. Look out for the big headings with a circle at each end – like the one at the top of this page. These headings tell you where a new topic starts.

Pages 4–11

Light and Shadows

Transparent, translucent and opaque materials; how light travels; shadow shapes; shadow games.

Pages 12–17

Big and Small Shadows

Size of shadows; shadows during the day or year; sundials; eclipses.

Pages 18–27

Seeing your Reflection

Shiny materials; mirror images.

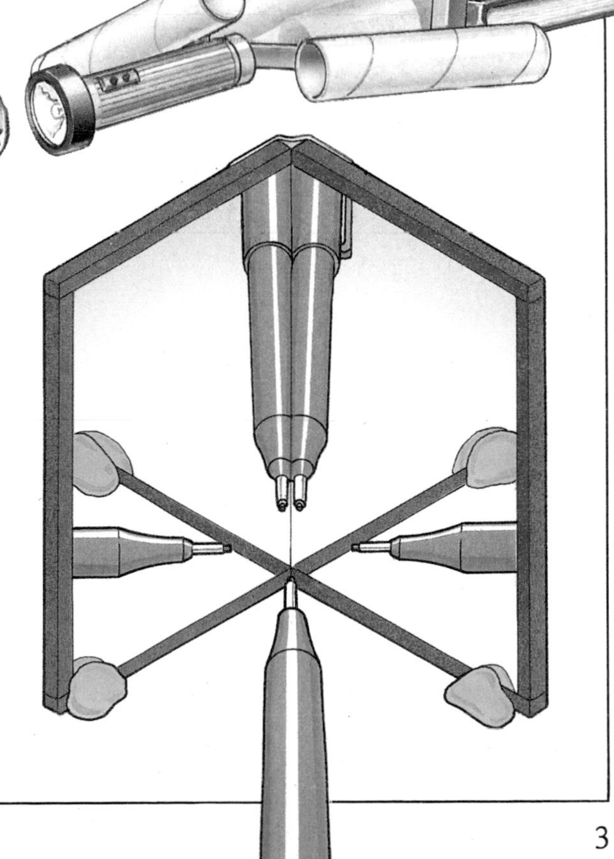

Pages 28–37

Bouncing Light

Angle of reflection; periscopes; kaleidoscopes.

Pages 38–39

Curved Mirrors

Reflections in spoons; fairground mirrors.

LIGHT AND SHADOWS

Make a collection of objects like the ones along the edges of these two pages. In a dark room, shine a torch on to each object. Which objects let the light through? Which objects keep out the light?

Some things let light go straight through them. You can see clearly through these things. They are said to be transparent. Clear glass and clean water are transparent.

Some things let light through but they scatter the light. If you look through these materials, everything looks blurred. These materials are said to be translucent. Frosted glass and tracing paper are translucent.

Many things do not let any light pass through them. You cannot see through these things. They are said to be opaque. Your body is opaque, so is this book.

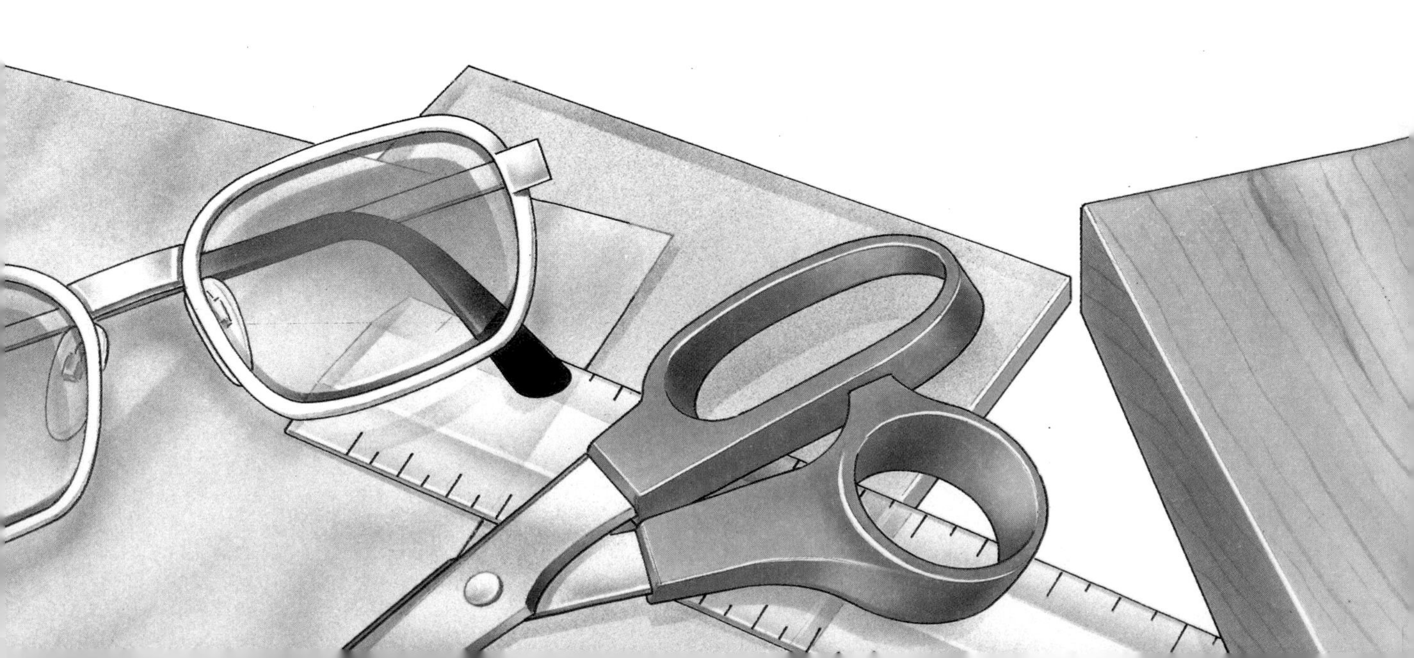

In this picture can you find a transparent material, a translucent material and an opaque material?

Going Straight

When light hits an opaque object, a dark area forms behind the object. This is called a shadow. Why doesn't light bend round objects and light up the shadow area? To find out, try this test.

1. Cut two pieces of card about 20 centimetres square.
2. To find the middle of the card, draw a line from each corner to the opposite corner. The point where the lines cross is the middle of the card.
3. Cut a hole in the middle of each piece of card.
4. Use modelling clay to fix the

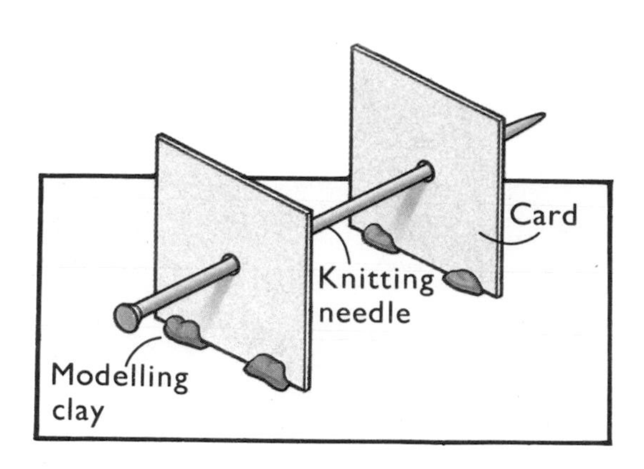

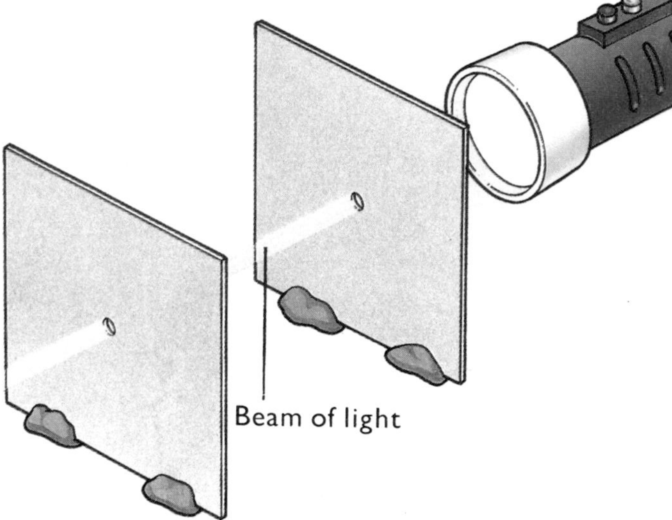

Make a Pinhole Camera

To investigate how light travels, try making a pinhole camera.

You will need:
An empty box, thick brown paper, tracing paper, a pin, scissors, sticky tape, black paint, charcoal or a thick black felt pen, a dark cloth or towel.

1. Cut both ends off the box.
2. Paint or colour the inside of the box black.
3. Tape a piece of brown paper over one end of the box.

cards upright about 30 centimetres apart. To line up the holes in a straight line, push a knitting needle through both holes.
5. Ask a friend to shine a light on to the first hole. You should see the light go through the second hole.
6. Now move the second card to one side so the holes are not in a straight line. What happens?

What happens
Light travels in straight lines and cannot bend around things. So when you move the second card out of line, the light cannot get through the second hole.

▲ Can you see the straight edges of these beams of sunlight?

4. Tape a piece of tracing paper over the other end.
5. Use the pin to make a small, round hole in the middle of the brown paper.
6. Cover your head and the tracing paper end of the box with the cloth or towel.
7. Point the camera at a window and look at the tracing paper from about 15 centimetres away. You should see an upside-down window.

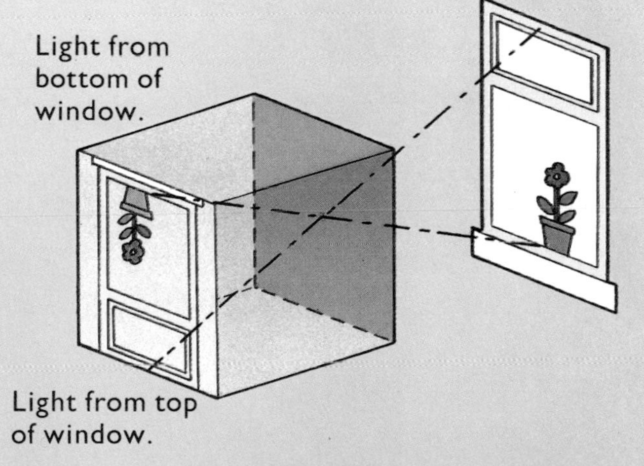

What happens
Light from the top of the window passes in a straight line through the pinhole to the bottom of the tracing paper. Light from the bottom of the window travels to the top of the tracing paper. So the picture you see is upside-down.

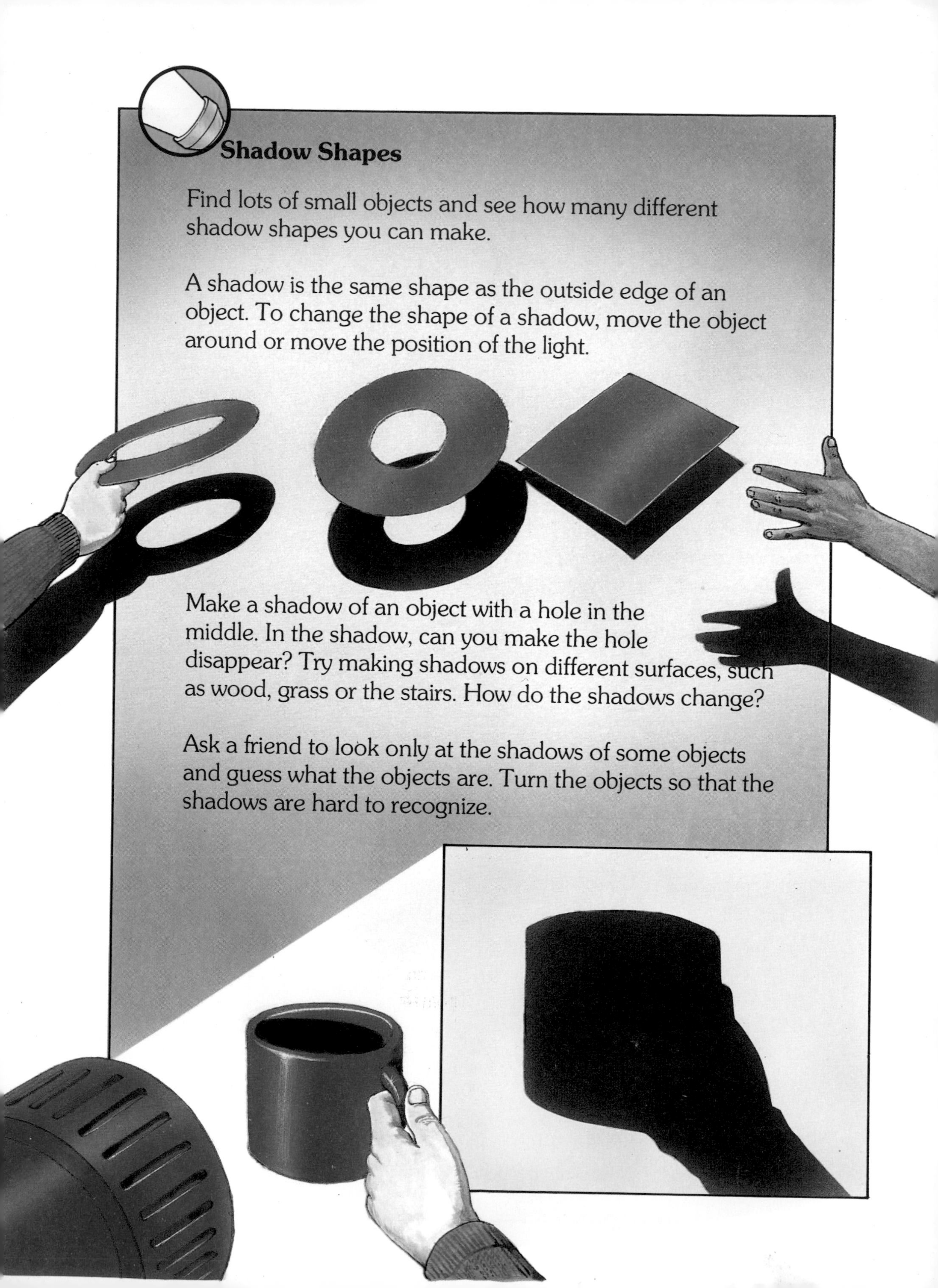

Shadow Shapes

Find lots of small objects and see how many different shadow shapes you can make.

A shadow is the same shape as the outside edge of an object. To change the shape of a shadow, move the object around or move the position of the light.

Make a shadow of an object with a hole in the middle. In the shadow, can you make the hole disappear? Try making shadows on different surfaces, such as wood, grass or the stairs. How do the shadows change?

Ask a friend to look only at the shadows of some objects and guess what the objects are. Turn the objects so that the shadows are hard to recognize.

Shadow Portraits

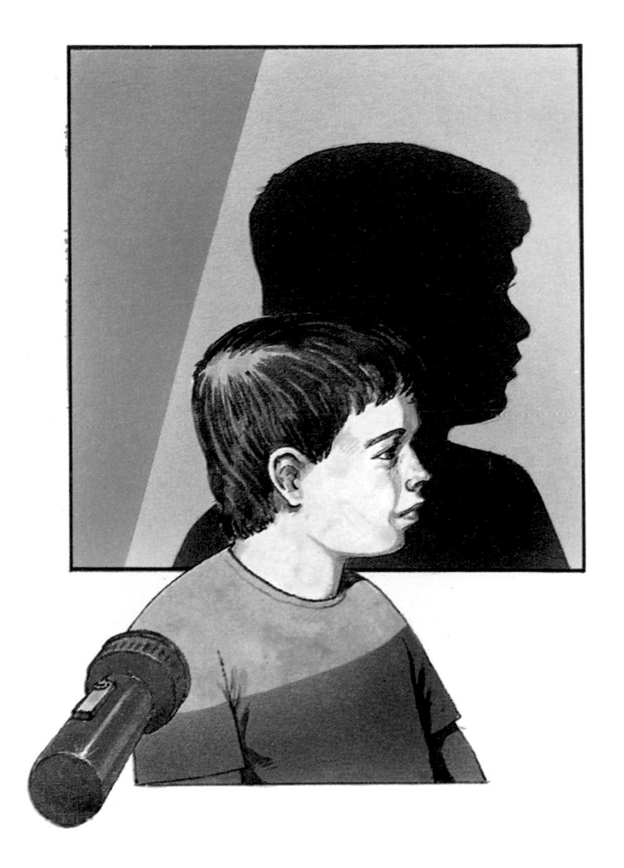

1. Ask a friend to sit sideways on a chair near a wall.
2. Use masking tape to fix a sheet of paper to the wall behind your friend's head.
3. Shine a torch so your friend's head casts a shadow on the paper.
4. Draw around the edge of the shadow with the pencil.
5. Paint inside the outline.

Animal Shadows

With your hands, you can make animal shadows. Here are some ideas. How many more can you discover?

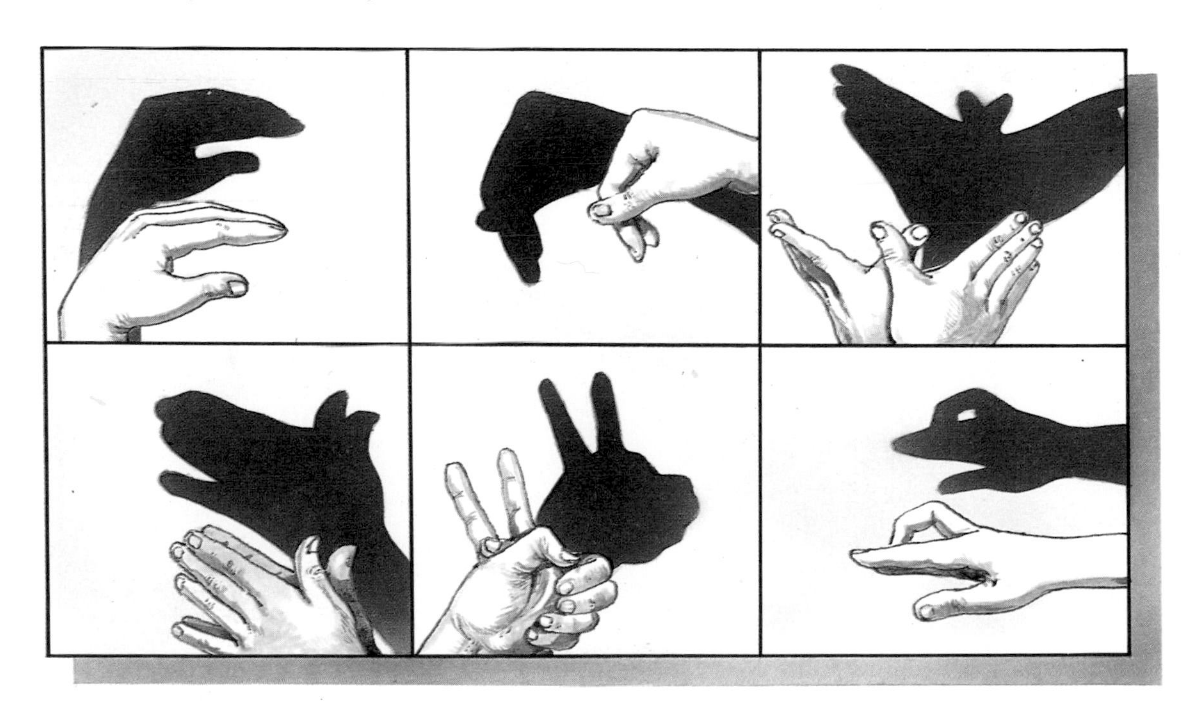

Shadow Games

Can you hide your shadow?
Can you escape from your shadow?

Can you shake hands with a friend's shadow if your hands are not touching?

Hide your shadow

Escape from your shadow

Shake hands

Can you catch your friend's shadow? If he is standing on the shadow of a tree or a building, then he is safe and cannot be caught. If you put a foot on his shadow, it's his turn to try and catch your shadow.

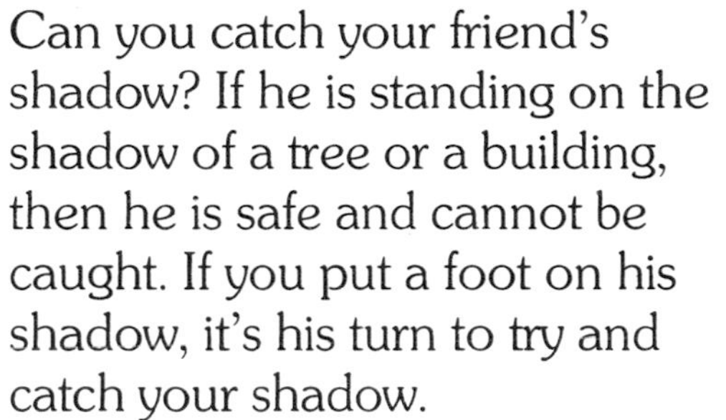

Can you make your shadow touch the top of the tree? Or reach into places where you cannot go?

Can you fit inside your friend's shadow?

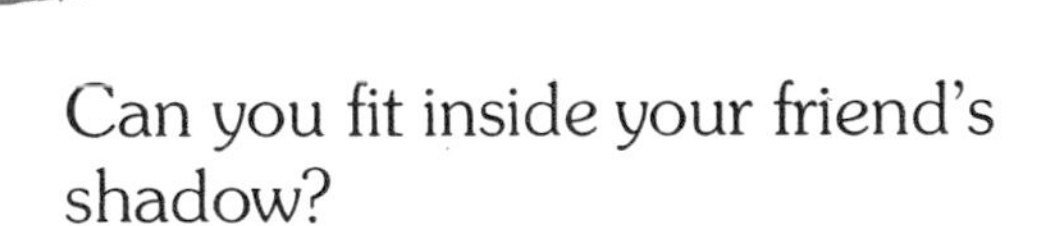

Can you jump on your friend's shadow?

BIG AND SMALL SHADOWS

Cut out a star shape from a piece of card and fix it to a pencil with modelling clay. Prop up a large piece of white paper on some books. In a dark room, shine a torch on to your star.

Hold the star near the torch.
How big is the shadow?

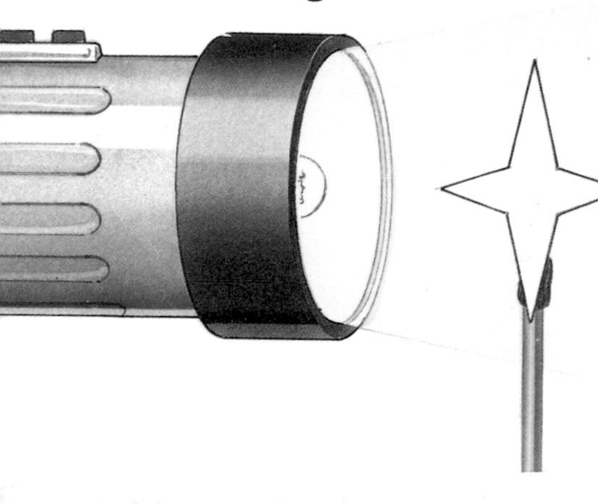

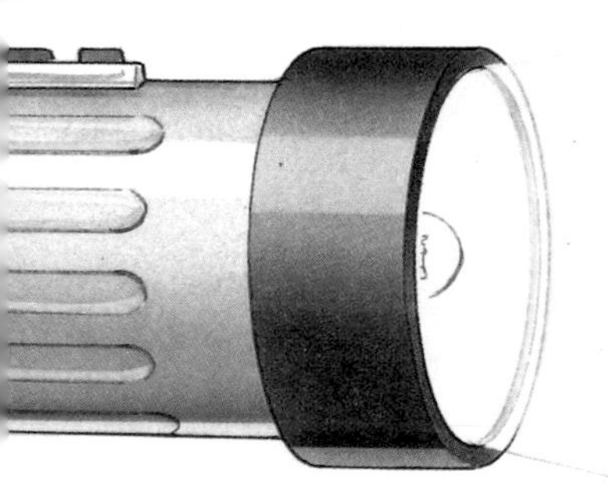

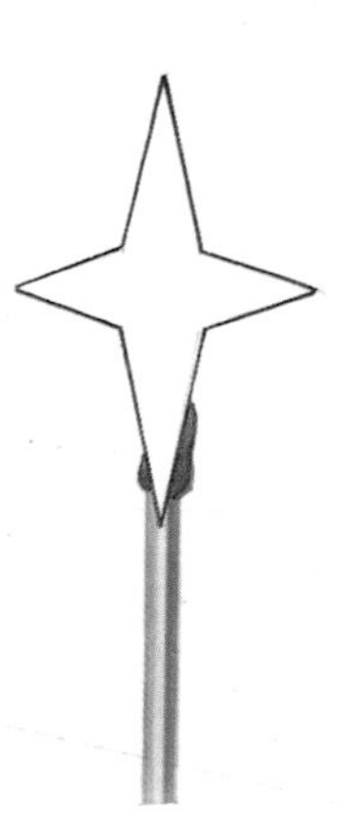

Now move the star farther away from the torch.
What happens to the shadow. Is it bigger or smaller?
What happens to the size of the shadow if you keep the star in one place and move the torch backwards and forwards?

What happens
When the star is near the light, it blocks out a lot of light, so the shadow is big. When the star is farther away from the light, it blocks out less light, so the shadow is smaller.

Long and short shadows

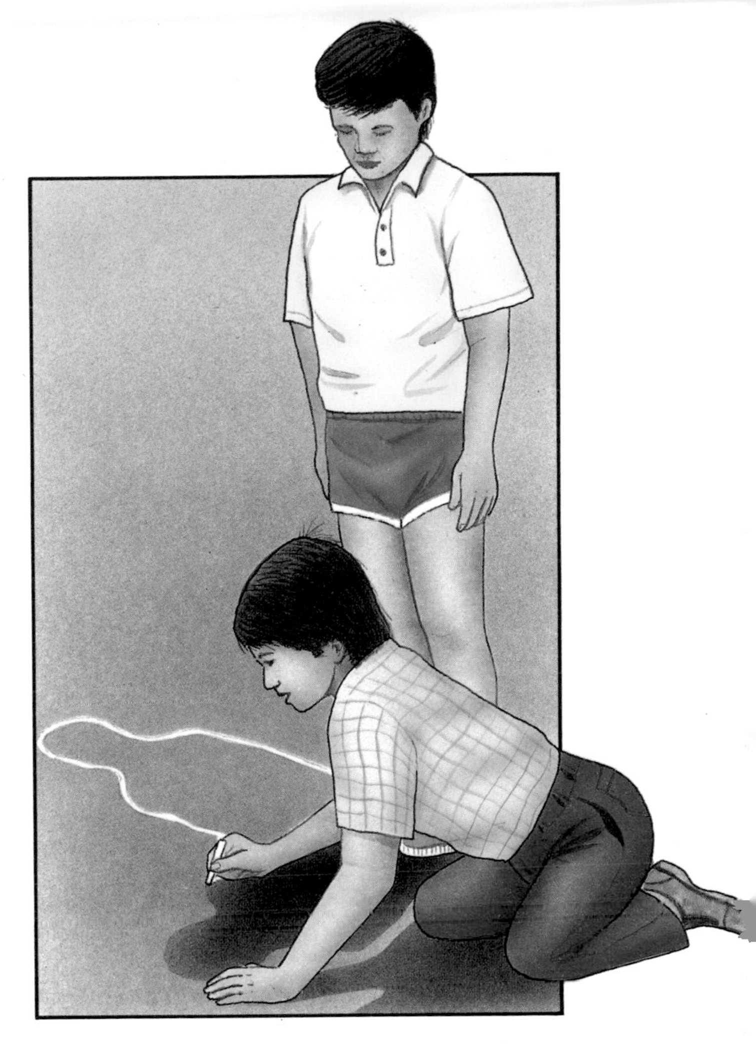

On a sunny day, find a safe area of concrete or asphalt – not out in the street. Ask if you can use chalk to draw around shadows. With a friend, stand in the same place in the early morning, at midday and in late afternoon. Take it in turns to stand still while one of you draws around the other's shadow.

Measure your shadows from head to toe. How does the size of the shadows change at different times of the day? Which direction do your shadows point at different times of the day – north, south, east or west?

Compare your shadows on a summer's day with your shadows on a day in winter. How are they different?

What happens
In the early morning or late afternoon, the Sun is low in the sky and shadows are long. At midday, the Sun is high above you and your shadows are shorter. In winter, the Sun is lower in the sky than it is in summer. So winter shadows are longer than summer shadows.

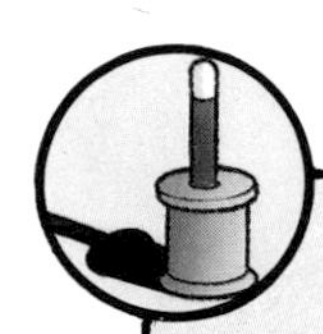

Make a Shadow Clock

You will need: a pencil or a short stick, a cotton reel, glue or modelling clay, a large piece of white paper.

You can use shadows to help you tell the time. Here's how to make a shadow clock.

1. Stand a pencil or a short stick in a cotton reel.
2. Use glue or modelling clay to fix the reel to a large piece of white paper.
3. On a sunny day, put your shadow clock out of doors where the Sun will shine on it.
4. Draw a line along the shadow of the stick and write the time at the end of the line.
5. Do the same thing every hour.

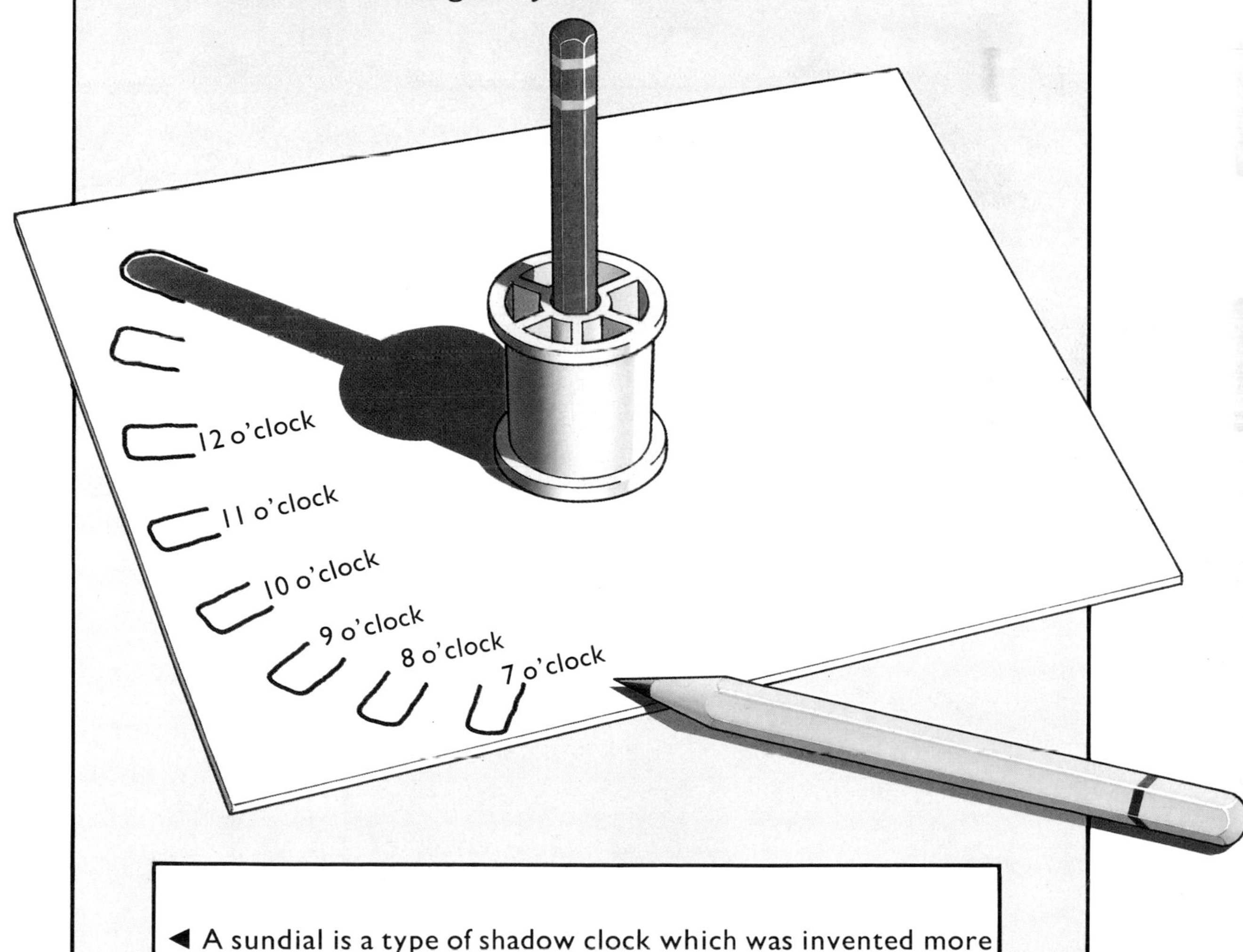

◄ A sundial is a type of shadow clock which was invented more than 3000 years ago, long before watches were made.

Shadows in Space

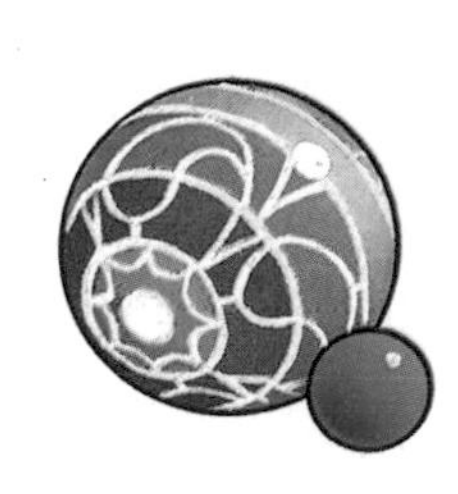

The Moon has no light of its own. It shines because it reflects light from the Sun. Sometimes, the Earth moves in a direct line between the Moon and the Sun and stops Sunlight from reaching the Moon. The Earth's shadow makes the Moon look very dark for a while. This is called a Lunar eclipse. 'Lunar' means to do with the Moon.
Try this experiment to see how a Lunar eclipse works.
Use a large beachball or a soccer ball as the Earth and a much smaller ball as the Moon. Stick a piece of string to the 'Moon' so you can hang it in front of the 'Earth'. Use a torch or a table lamp as the Sun.

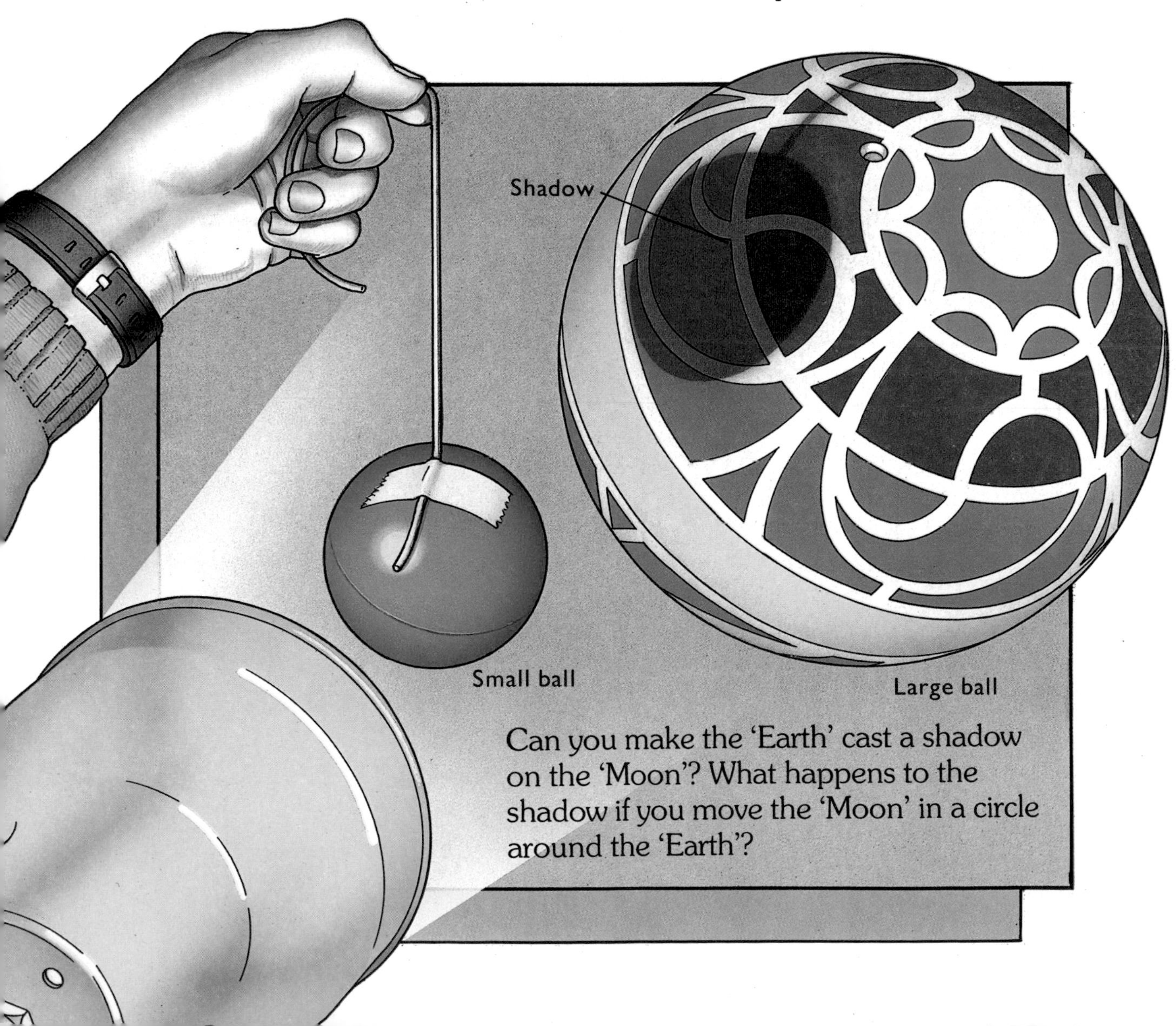

Can you make the 'Earth' cast a shadow on the 'Moon'? What happens to the shadow if you move the 'Moon' in a circle around the 'Earth'?

▲ This photograph shows what happens when the Moon passes in front of the Sun. It stops Sunlight from reaching the Earth and makes some places on Earth dark during the daytime. This is called a Solar eclipse – 'Solar' means to do with the Sun.

For a Solar eclipse to happen, the Moon, the Sun and the Earth all have to be in a straight line. This does not happen very often.

SEEING YOUR REFLECTION

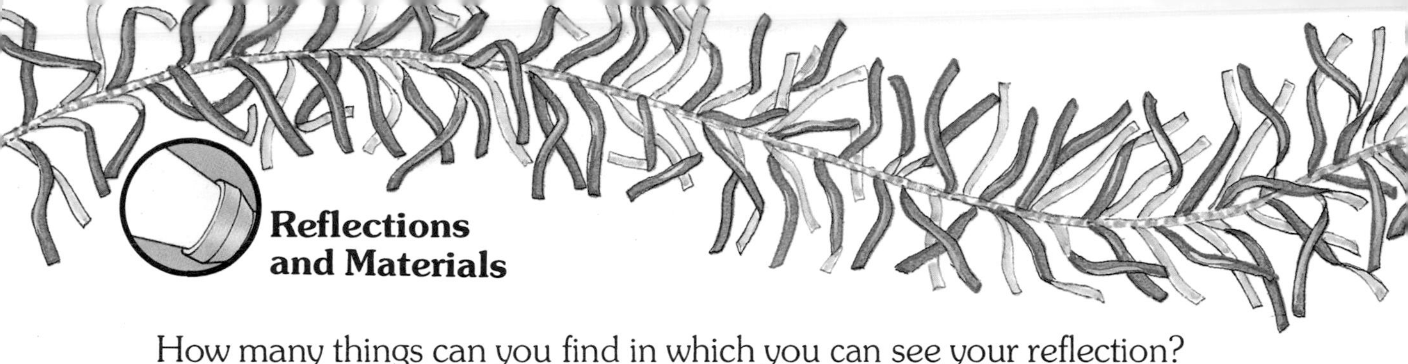

Reflections and Materials

How many things can you find in which you can see your reflection? Look for your reflection out of doors in shop windows, car bonnets and buildings. Can you see your reflection in water? What happens to your reflection if the water is moving?

In a dark room, shine a torch on to different materials such as tin foil, brick, wood, plastic, cloth and metal. Which materials give the best reflections?

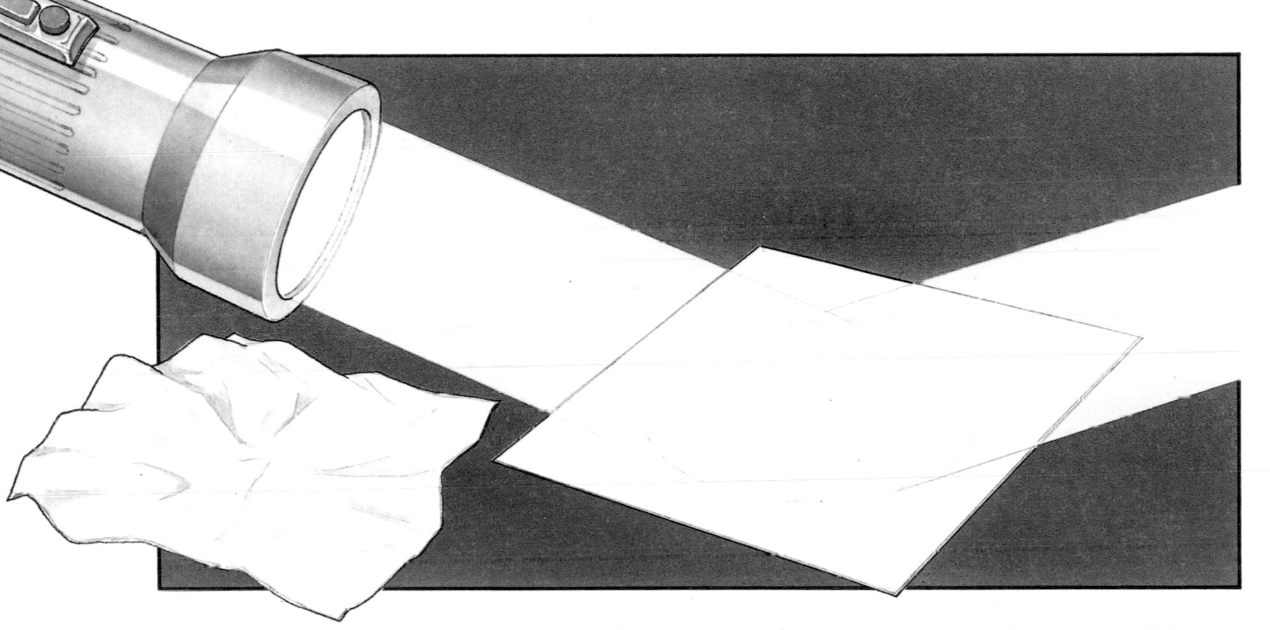

Compare a smooth piece of foil with a wrinkled piece of foil. Can you see your face in both pieces of foil?

Make a collection of shiny materials. You could sort your collection into groups according to the materials the objects are made of – metal or non-metal for example.

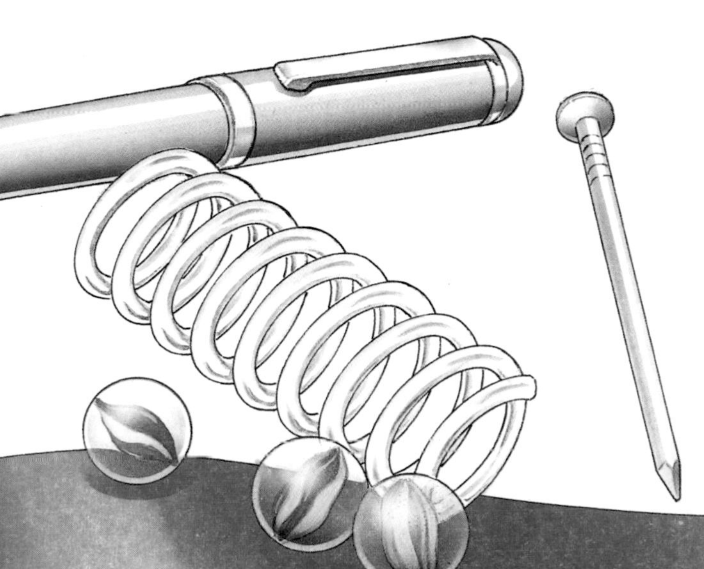

Mirror, Mirror

Smooth, shiny surfaces produce the best reflections. This is why mirrors are made of a flat sheet of glass with a thin layer of shiny metal, such as silver or aluminium behind the glass.

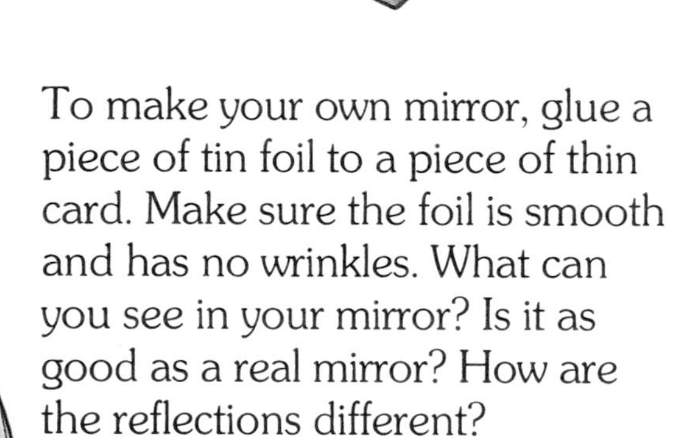

To make your own mirror, glue a piece of tin foil to a piece of thin card. Make sure the foil is smooth and has no wrinkles. What can you see in your mirror? Is it as good as a real mirror? How are the reflections different?

How many different mirrors can you find at home, at school, along the street or in the shops? What shapes and sizes are they? How are they used?

▶ Dancers and actors would find it difficult to put on their make-up properly without a mirror.

Look in the Mirror

Look at your reflection in a mirror. Does your reflection do what you do?
Touch your right ear. Which ear does your reflection touch?

Your reflection seems to be touching its left ear. The reflection is the wrong way round and does not show you as other people see you.

A reflection does not come from the surface of a mirror. It seems to be behind the mirror. Try measuring the distance between you and the mirror. You will find that your reflection appears to be the same distance behind the mirror as you are in front of it.

Mirror Maze

To find out more about the reflections in mirrors, play this game with your friends.

1. Draw a large star shape on a piece of paper.
2. Draw a second star a few millimetres outside the first one.
3. Use some books to prop a large mirror upright on a table. Or ask a friend to hold the mirror for you.
4. Hold a large book in front of your star maze so you can see the reflection but not the actual drawing.
5. Now, looking only in the mirror, try to draw around the maze without touching either of the lines.
6. How fast can you do this? Can your friends beat your fastest time? If they touch the lines, they must start again.

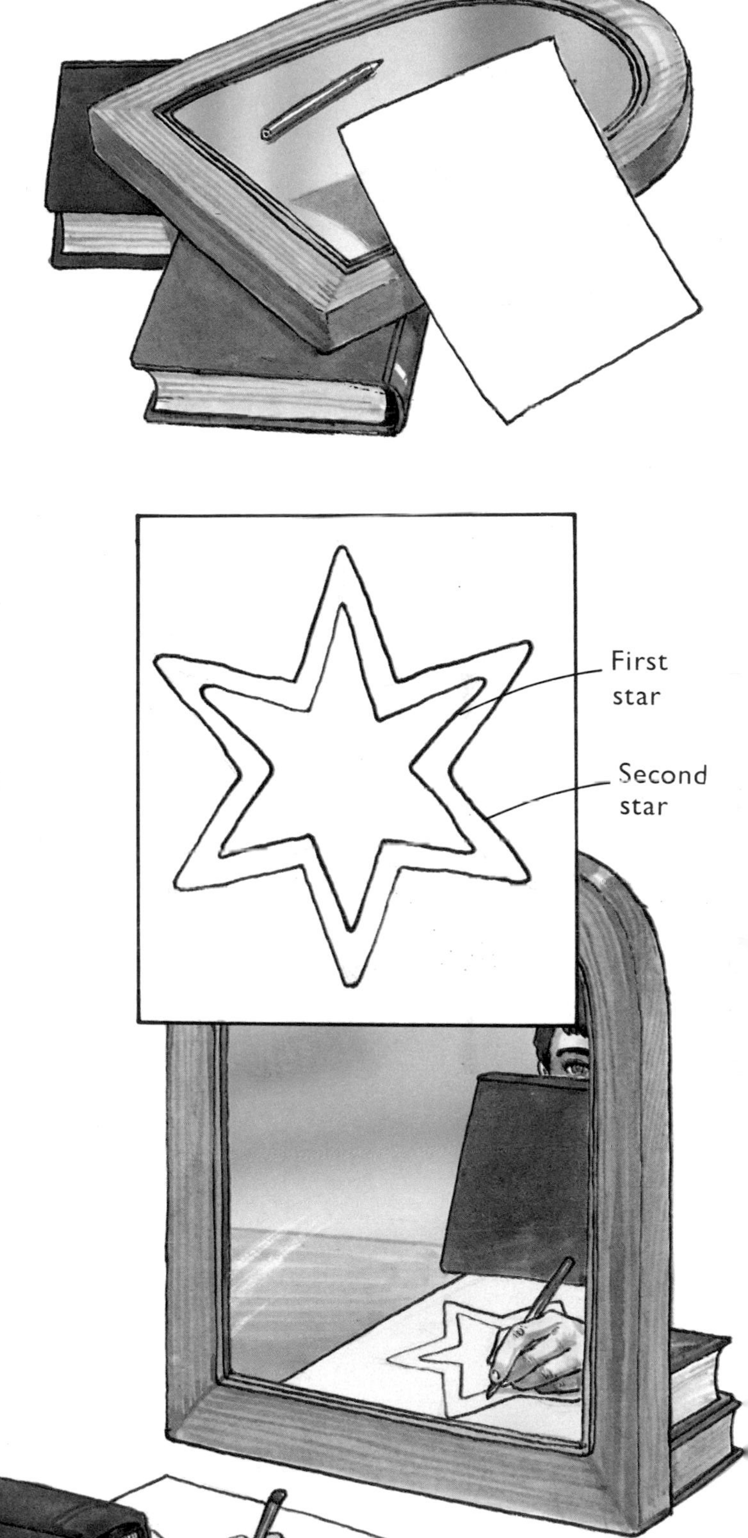

Mirror Writing

Write your name in big letters on a piece of paper. Hold the paper in front of a mirror. Can you read your name? What has happened to the letters?

Many ambulances have the word 'AMBULANCE' written back to front on the outside of the vehicle. This is so that other drivers see the word the right way round when they look in their rear-view mirrors. It helps drivers to spot an ambulance coming up behind them.

Hold up this book behind your head and look at the picture below in a mirror. In your mirror, is the word the right way round?

Secret Mirror Code

You can use a mirror to invent your own secret code. Only someone with a mirror will be able to read your messages.

The trick is to write your messages while looking only in the mirror. Write each letter slowly so it looks correct in the mirror. Don't look down at the piece of paper.

When you have finished your message, it will look strange because the letters are back to front. Some of the letters will not change. Which ones are they?

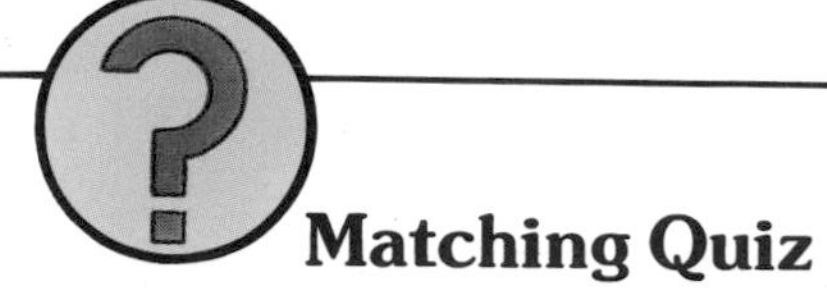

Matching Quiz

Some things can be divided by an imaginary line into two parts that look exactly the same. These things are called symmetrical objects. You can find out if something is symmetrical by putting a mirror along the dividing line. If the object looks the same in the mirror, it has a kind of symmetry.

See if you can guess which of the objects below are symmetrical. Test them by holding a mirror against each picture.

Funny Faces

Is your face symmetrical? Find a photograph of yourself which shows your whole face from the front. Hold a mirror down the middle of the photograph. Look at both sides of your face. Does your face look strange?

Mirror Painting

To make a symmetrical painting, **you will need:** newspaper, plain paper, poster paints, jar of water, paintbrush, apron.

1. Put on the apron and lay some newspaper on a table or on the floor so you won't make a mess.
2. Mix up several different colours with the poster paints. Make each colour fairly thick.
3. Use the brush to drop or brush paint on to one side of the paper.
4. Fold the paper over and press down to smooth out the paint.

What happens

When you open out the paper, you will have a painting that is the same on both sides, just as if you were looking in a mirror.

BOUNCING LIGHT

Reflections are caused by light bouncing off things. When this light is reflected into our eyes, we are able to see things.

Catch the Light

Use a small mirror to reflect a spot of light from the Sun or a lamp on to a wall. If you turn the mirror a little, the light spot will move too. Ask a friend to make another light spot on the same wall. Can your friend touch your light spot with the one they have made?

Bouncing Back

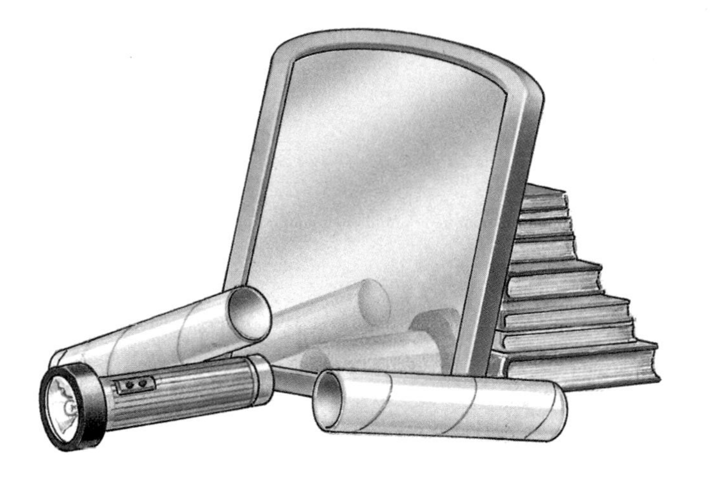

When light hits a smooth surface, it always bounces back at a matching angle. To see how this works, try this test.

You will need: a large mirror, two cardboard tubes, a torch, some books.

1. Use the books to prop the mirror upright.
2. Hold one tube at an angle with the end touching the mirror.
3. Ask a friend to hold the second tube at a matching angle.

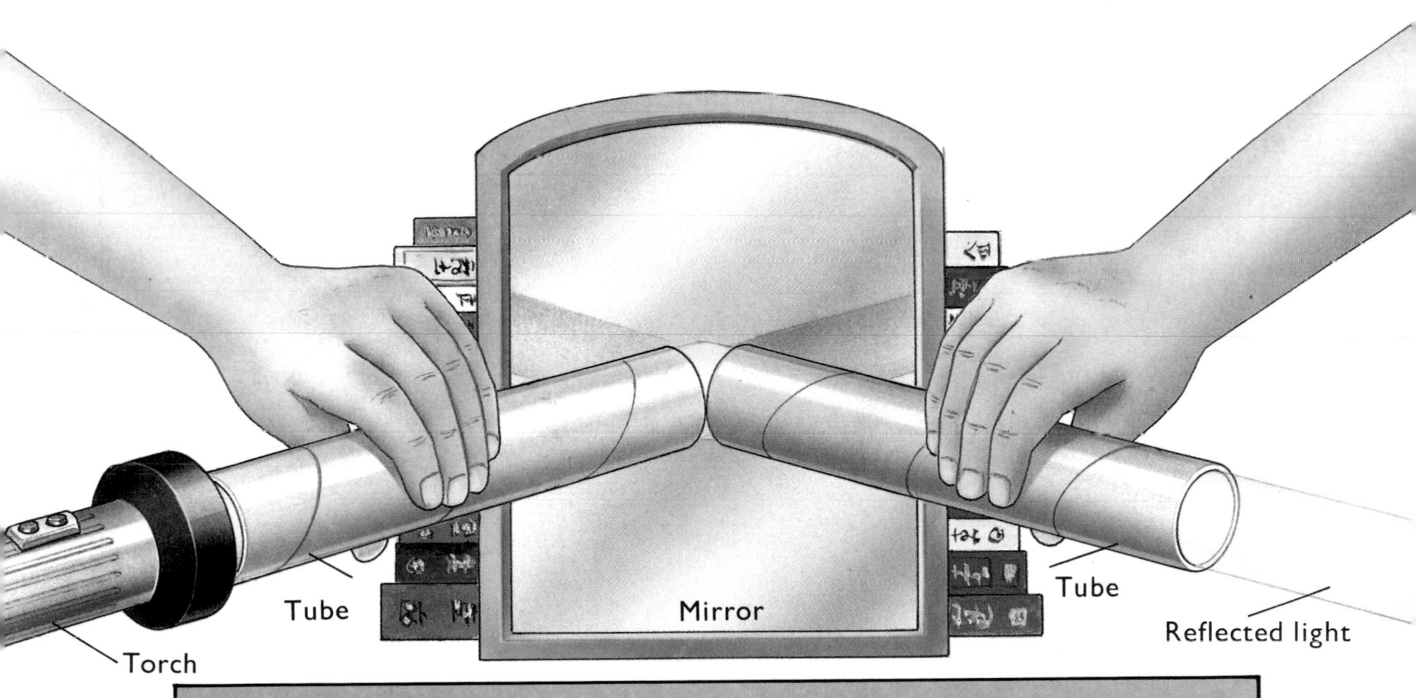

What happens

When the tubes are at the correct angle, the light will bounce off the mirror and down to the end of the second tube. If your friend holds their hand at the end of the tube, they will see a circle of reflected light.

On a rough surface, light is not reflected like this. It is scattered back in several different directions.

▲ These rows of mirrors reflect the Sun's rays on to the tower, which is a Solar heat collector. This concentrates a lot of the Sun's energy in one place. The mirrors turn to follow the path of the Sun as it moves across the sky. Solar power is used to generate electricity, to heat and cool buildings and to power small objects such as watches and calculators. Using energy from the Sun does not use up any of the Earth's resources or cause pollution.

See-through Brick

Can you make a light go through a brick?

You will need: a brick, a torch, a piece of cardboard, four small mirrors, modelling clay, scissors.

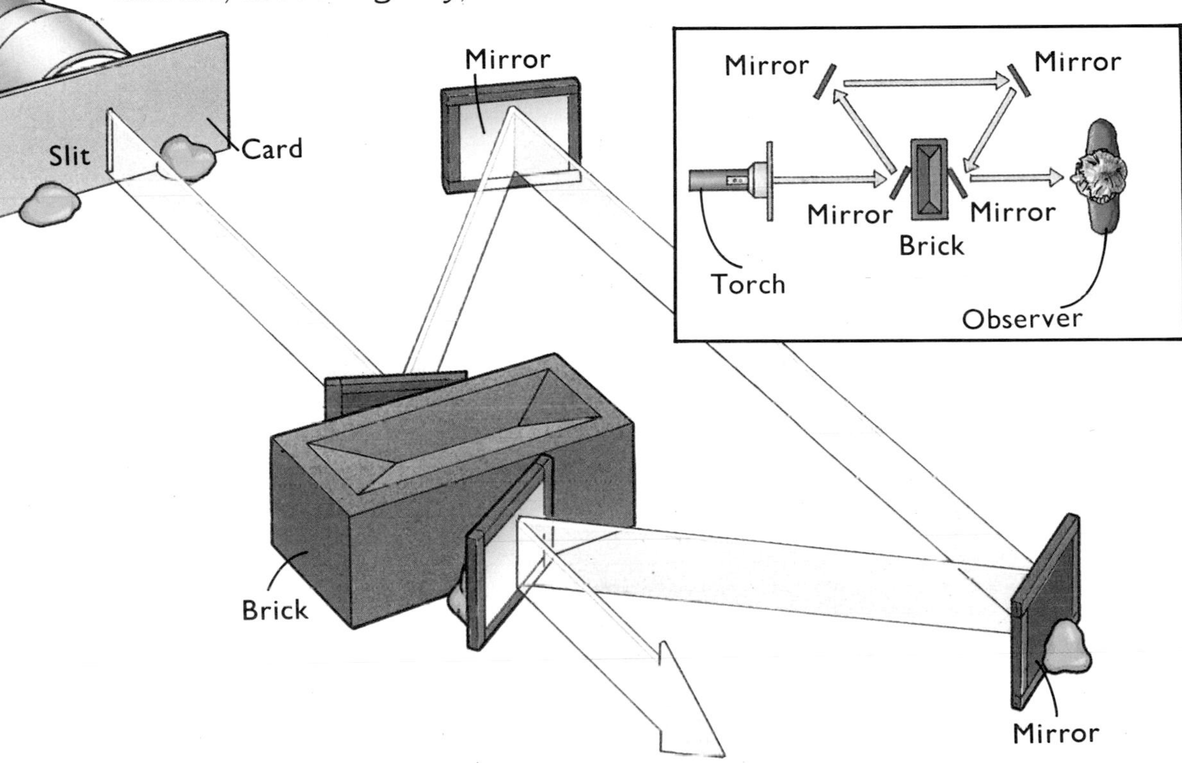

1. Cut a narrow slit about six centimetres deep in the middle of one edge of the piece of cardboard.
2. Use some modelling clay to prop the cardboard upright.
3. Place the brick a little way in front of the card.
4. Use the modelling clay to fix the mirrors to match those in the picture.
5. Shine the torch through the slit.

What happens

The light bounces from one mirror to another and looks as if it is coming straight through the brick.

Multiplying Mirrors

When two mirrors are held together at an angle, the light bounces to and fro between the mirrors. This means you can see more than one reflection of an object.

Tape two mirrors together along one of the long sides. Use some modelling clay to stand the mirrors upright at a wide angle. Put a small object in front of the mirrors. How many reflections can you see?

Now move the mirrors closer together. Count how many reflections of the object you can see now.

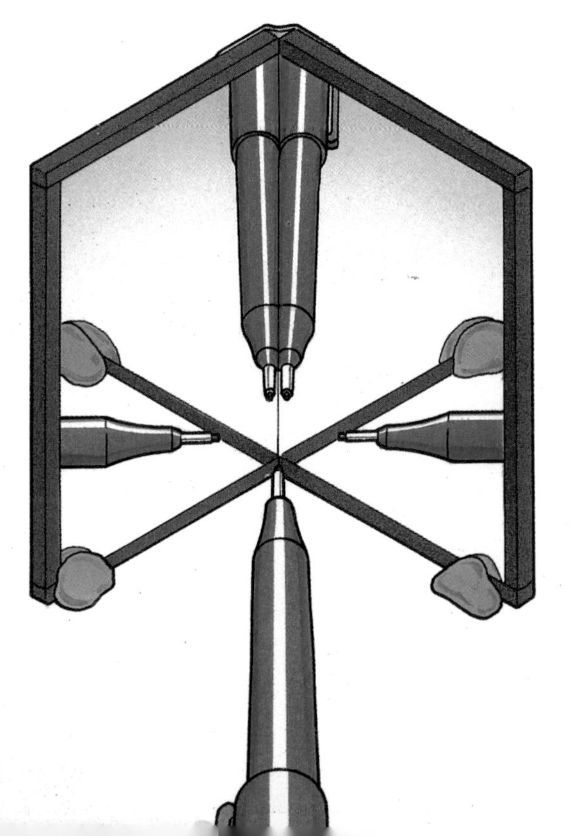

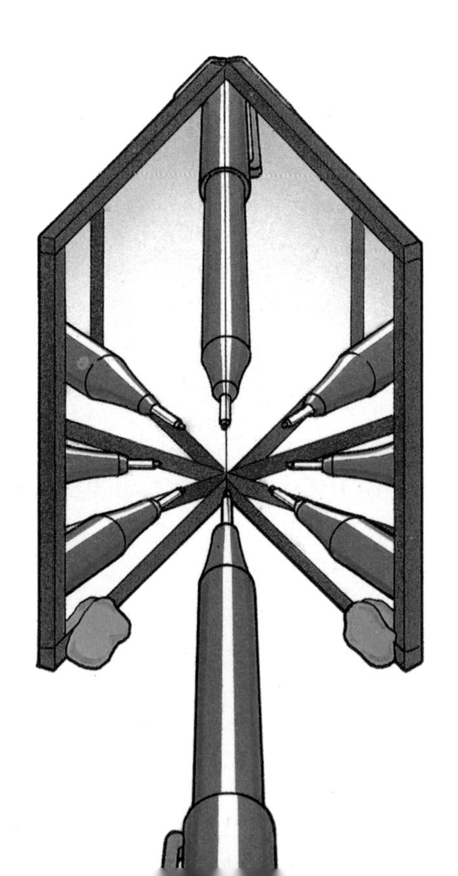

Your Other Face

You can use two mirrors to see yourself as other people see you. To see how this works, you need to make one side of your face look different from the other side. You could use face paints to do this.

Look in one mirror and remember where the painted side of your face appears on your reflection. Then hold two mirrors facing each other at an angle. Move the mirrors until you can see your whole face at the point where the two mirrors join. The painted side of your face will now be on the other side of your reflection.

What happens
With the mirrors at an angle, the reflection of the left side of your face bounces across to the right hand mirror. And the reflection of the right side of your face bounces across to the left hand mirror. When other people look straight at you, they see your face this way round.

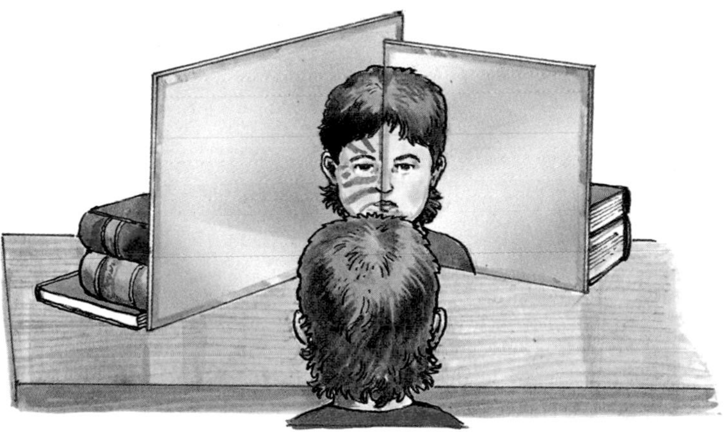

How Many Reflections?

Hold a small mirror facing a large mirror so the small mirror is just in front of your nose. As you look over the top of the small mirror, you will be able to see lots and lots of reflections stretching away into the distance. How many reflections can you see? Are all the reflections the same size?

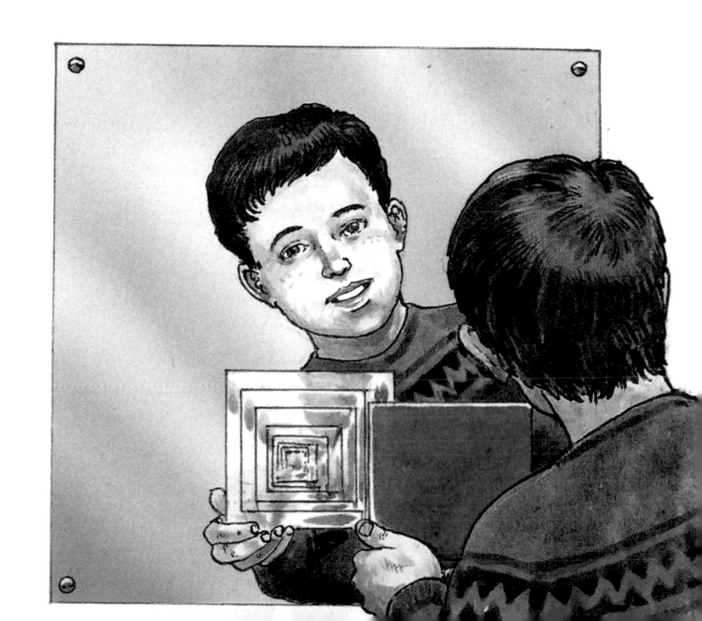

Make a Periscope

Have you ever had your view blocked by a crowd of people? By making a periscope you will be able to see over their heads so you won't miss anything. You can also use a periscope to look round a corner or over a wall without being seen.

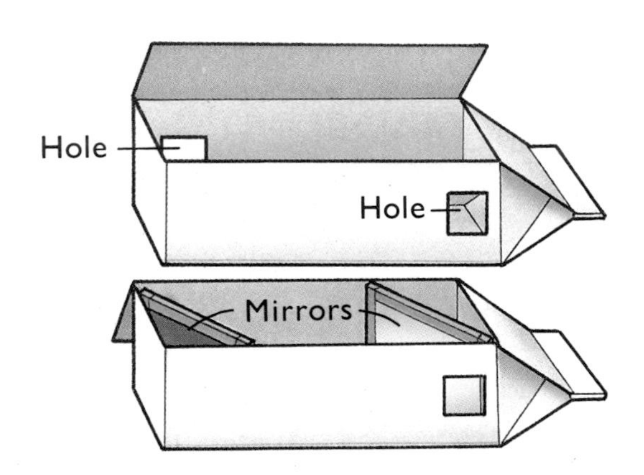

You will need: a large empty milk or juice carton, two small mirrors, sticky tape, scissors.

1. On one side of the carton, cut three of the edges to make a lid you can lift up.
2. Cut two holes in opposite sides of the carton, to match the picture.
3. Tape the two mirrors inside the carton. The mirrors should be facing each other at the same angle.
4. Tape down the lid.
5. Hold the periscope upright and look into the bottom mirror.

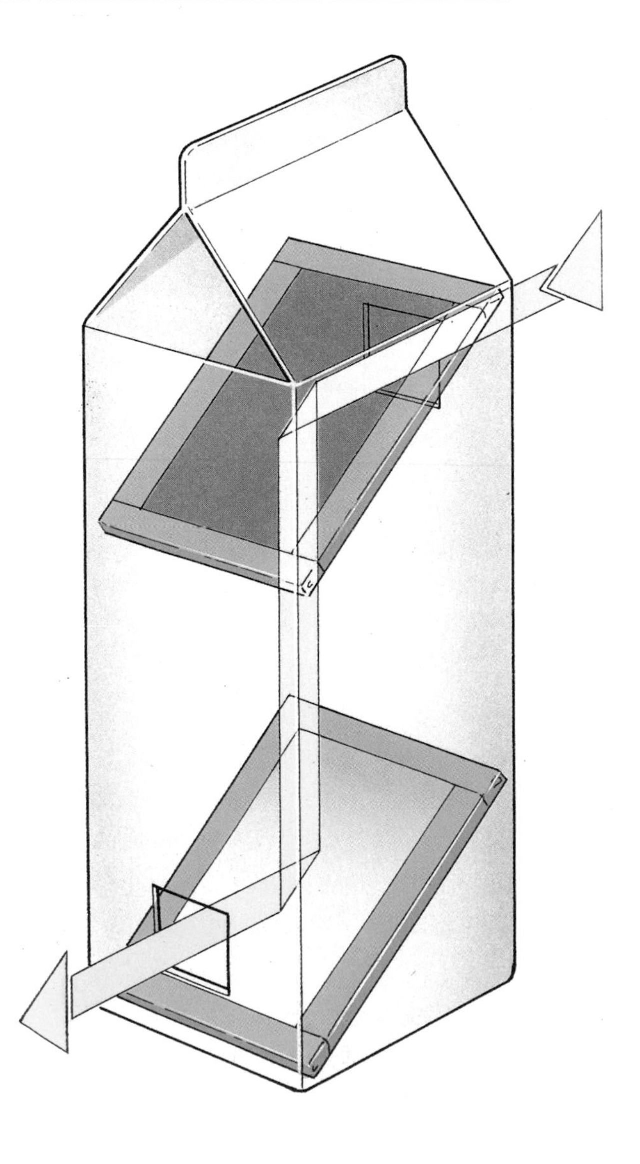

What happens

The light bounces from one mirror to the other so you can see over people's heads or round corners.

Using periscopes, even people at the back of a large crowd can see what is happening at the front.

Make a Kaleidoscope

The word 'kaleidoscope' means 'beautiful form to look at'. If you make a kaleidoscope, you will be able to see lots of beautiful symmetrical patterns.

You will need: stiff cardboard, a pencil, scissors, black paper or a thick black felt pen, tin foil, glue, clear plastic, tracing paper, sticky tape, small coloured shapes or beads.

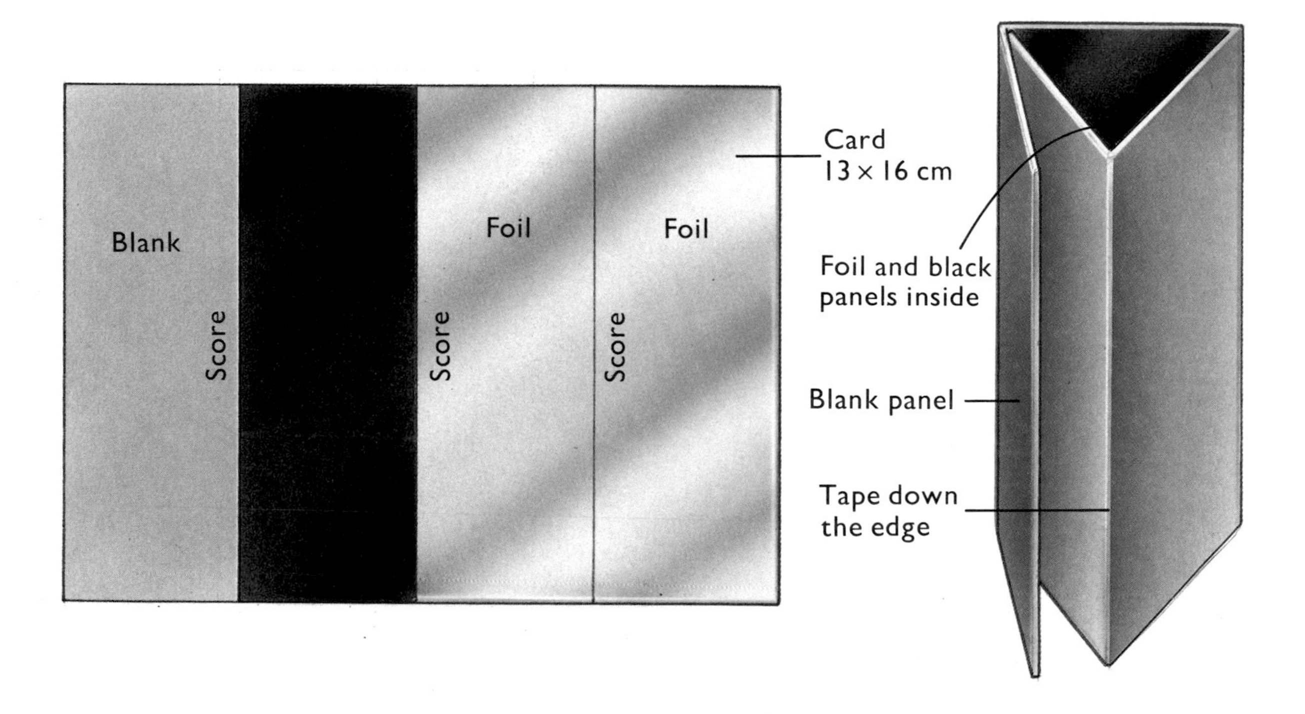

1. Cut out a piece of cardboard about 13 centimetres by 16 centimetres.
2. With the pencil, divide the card into four equal strips. Each strip should be four centimetres wide.
3. Ask an adult to help you score the lines so the card is easier to fold.
4. Stick foil over two of the panels. Make sure it is as smooth as possible.
5. Stick black paper over the third panel or colour it black.
6. Leave the fourth panel blank.
7. Fold the card to make a triangular shape and tape the side to hold it in place.

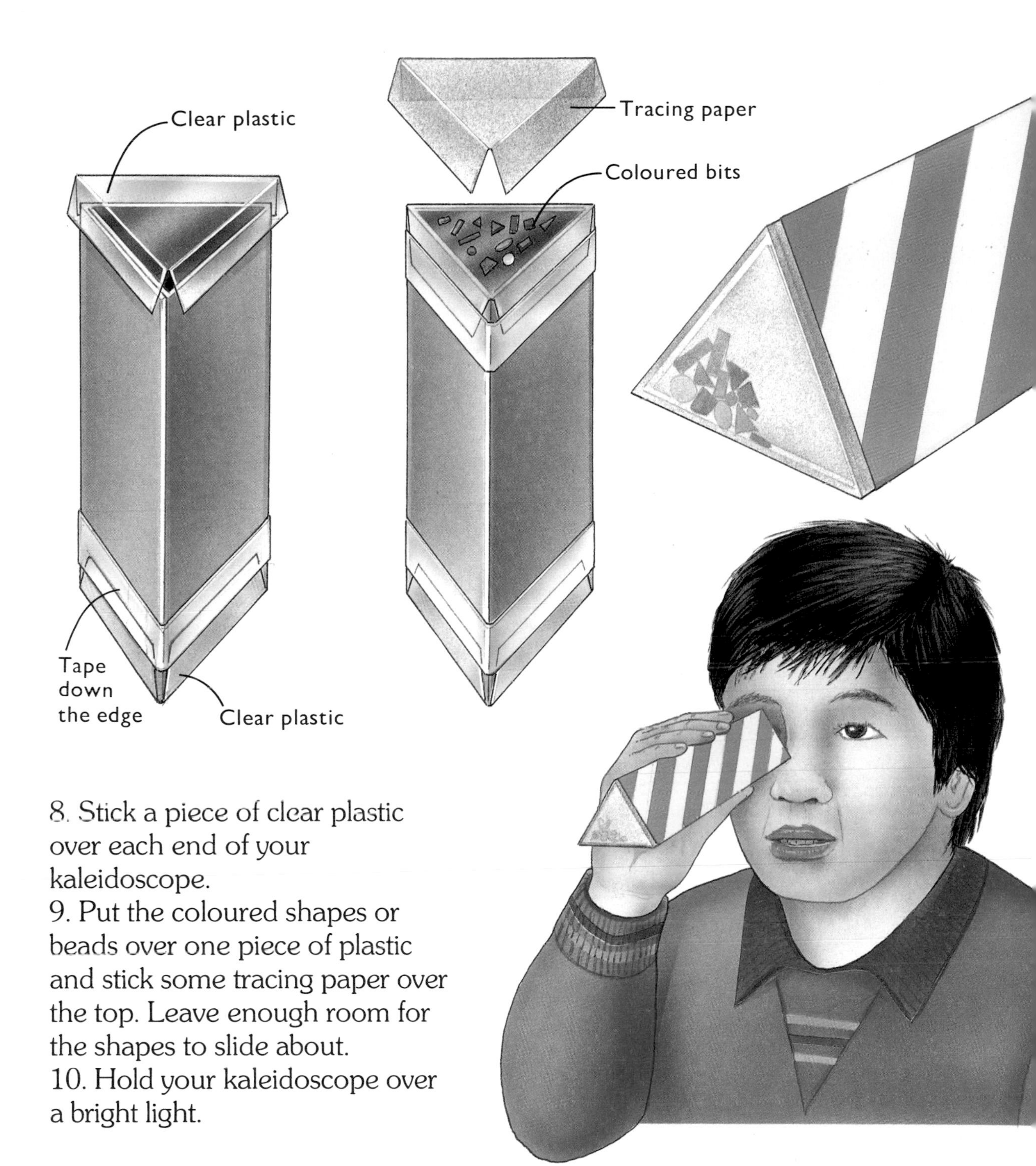

8. Stick a piece of clear plastic over each end of your kaleidoscope.
9. Put the coloured shapes or beads over one piece of plastic and stick some tracing paper over the top. Leave enough room for the shapes to slide about.
10. Hold your kaleidoscope over a bright light.

What happens
The light bounces to and fro between the foil mirrors. The reflections of the coloured shapes or beads make interesting patterns. To change the pattern, shake your kaleidoscope so the shapes or beads move into new positions.

CURVED MIRRORS

Curved mirrors change the size and shape of things reflected in them. Look at your reflection in the curved side of a shiny tin or a saucepan. What do you look like?

Now try a spoon. The back of a spoon curves outwards. This sort of curved mirror makes you look smaller.

What happens to your reflection in the front of a spoon?

► The surfaces of these curved mirrors curve both inwards and outwards. Look what it does to these reflections!

INDEX

Page numbers in *italics* refer to illustrations or where illustrations and text occur on the same page.

brick, see-through *31*

camera, pinhole *6–7*

face *27, 33*

glass 4, *5*

image inverted *6–7*

kaleidoscope *36–37*

light 4
lunar eclipse *16*

mirror *20–36*
  code *25*
  curved *38–39*
  maze *23*
  multiplying *32*
  painting *27*
  writing *24*
moon *16–17*

opaque 4, *5*, 6

periscope *34–35*

reflection *18–19, 22–23, 28–29*

shadow *6, 8–9*
  animals *9*
  games *10–11*
  portrait *9*
  size *12–13*

shadow clock *15*
solar eclipse *17*
solar heat collector *30*
solar power *30*
sundial *14*
sunlight *7*, 16–17
symmetry 26–27, 36–37

tracing paper 4
translucent 4, *5*
transport 4, *5*

water 4

Adviser: Robert Pressling
Designer: Ben White
Editors: Nicola Barber and Annabel Warburg
Picture Research: Elaine Willis

The publishers wish to thank the following artists for contributing to this book:
Peter Bull: page headings; Peter Dennis (Linda Rogers Associates): pp. 22–27, 33; Kuo Kang Chen: pp. 4/5, 12/13, 18/19, 28/29, 36/37; John Scorey: pp. 6–11, 15, 16, 20, 30–31, 32, 34, 38.

The publishers also wish to thank the following for providing photographs for this book:
Michael Holford; 7, 17, 18, 21, 35 Hutchison Library; 39 Spectrum; 5, 30 ZEFA.

Kingfisher Books, Grisewood and Dempsey Ltd,
Elsley House, 24–30 Great Titchfield Street,
London W1P 7AD

First published in 1990 by Kingfisher Books

British Library Cataloguing in Publication Data
Taylor, Barbara
  Shadows and reflections.
  1. Shadows. Shapes
  I. Title II. Series
  535

ISBN 0-86272-527-5

Phototypeset by Southern Positives and Negatives (SPAN), Lingfield, Surrey
Colour Separations by Scantrans pte Ltd, Singapore and Newselé Litho, Milan
Printed in Spain